God's Code

Male & Female

DNA

By Rasheed L. Muhammad

Copyright©2018

Contents

Forward 3
   Male Testicles Produce x Female Sex Gene 4
   God's Genetic Pattern In Man and Woman 7
   Nature of Human Being Is Righteousness 11
   Cultural Psychology Of God 12
   The Wobble In Nature 19
   Number 8 28
   God Speaks With Numbers 29
   Let There Be Light 32
   Appendix 1 35
   Brain Creation 35
   Appendix 2 37
   NASA Solves Mysteries About Wobbling Earth 37
   Appendix 3 38
   DNA and RNA Masculine and Feminine Principles 38
   Appendix 4 39
   Analemma Sun Pattern 39
Glossary 40
   DNA Attributes Are Attributes of Alllah 44

Forward

The word God is used by many religions to refer to an invoked or summoned one; Theos or Deus. Other Biblical scholars derive the word God from Babylonian gad (gawd) which was their —god of fortune.[1]

Many ancient African languages associate God with the sky, "Spirit Possession", and heaven above:

> *"The ancient Egyptians, and other African nations, associated their gods with the purity and the life causing essence of water: Egyptian ntr —natron| (cleansing agent) ntr —God| (unseen fructifying agent) (Coptic noute) Twi ntoro —spirit of patrilineage| Yorùbá ntori —because| Lugbara adro —guardian spirit| Adro —God| (also the whirlwind found in rivers) Mbuti Ndura —God| (<of the rainforest) ciLuba Ndele —divine, begetter, Ancestor| Gurma Untenu —God| Gurmantche Untenu —God| Fulani Ntori —God| Masai Naiteru —God| Kwasio Nture —sacred| Mombutu Noro —God| Ewe Tre —clan spirit, fetish| Ijo Toru —river| (Egyptian i-trw —river|) Tonga Tilo —blue sky, God| (from which the rains fall) Amarigna AnäTära —pure| Wolof Twr —protecting god, totem| Twr —libation| (Egyptian twr —libation|)"*[2]

On the other hand, to comprehend God as a Being which means, *reality itself: existence, change, properties, space, time and causality* is to add form, mass and movement to a reality that is not formless. So, this book illustrates God's power and force from gene expression (thought) to nucleus, to single cell to organ, to embryo to male and female. Elucidating how male and female represent an extension of God's DNA from the y to x chromosome is the aim of this 44 page book.

---

[1] http://www.ccg.org/english/s/p220.html

[2] file:///C:/Users/dhras/Downloads/AfricanOriginsoftheWordGod%20(1).pdf

Male Testicles Produce x Female Sex Gene

The mind of a nation is hidden deep within the nucleus of the DNA cell composed of 4 basic chemical compounds and light (bio-photons). Your and my genetic ancestors also travel or live through our DNA chromosomes in thought, physical traits, and genetic codons. In fact, the population founder of any given ***living being*** will appear on all lines of the genealogy [germ cell] if one could trace their genealogy back in time. What you'll come to realize is human life began with a black male. Then came forth a black female — mother of civilization. By this we mean, in male spermatogenesis both sex determination genes (x and y) originate deep inside male testicles.[3] (See Diagram 1)

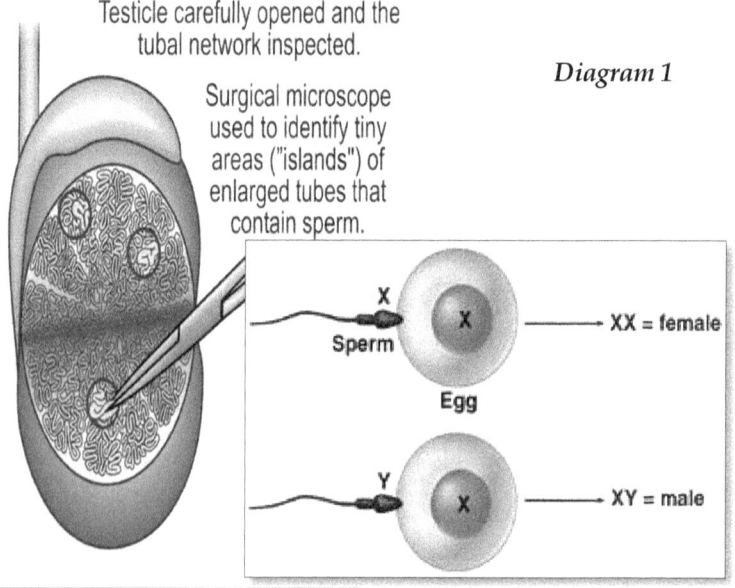

Diagram 1

---

[3] http://dev.biologists.org/content/134/10/1823

> *"In humans...the father determines the sex of the child. In the XY sex-determination system, the female-provided ovum contributes an X chromosome and the male-provided sperm contributes either an X chromosome or a Y chromosome, resulting in female (XX) or male (XY) offspring, respectively."*[4]

After a sperm cell matures, it travels into a female egg cell, which only produces her own "x" sex gene. Both male and female genes contain 23 chromosomes for a total of 46 chromosomes. This contribution by both is an extension of Gods DNA formula. Upon fertilization, the women co-creates an offspring with many ancestral traits preserved in the genetic book of inheritance (DNA) as shown below.

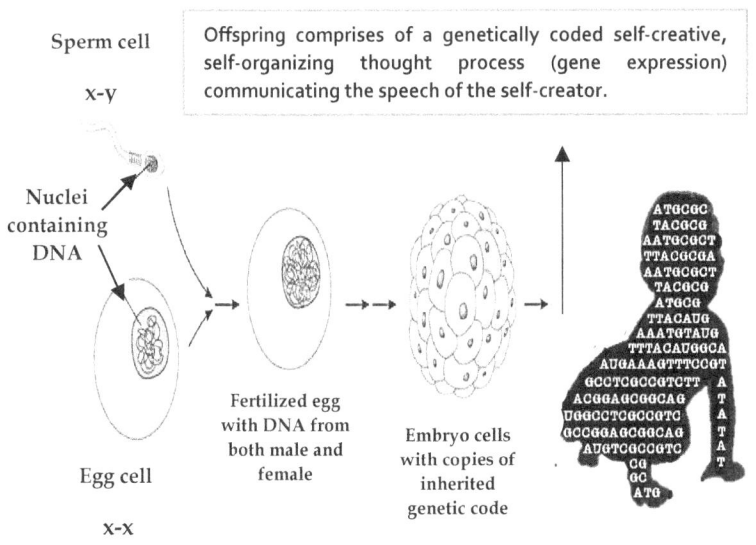

---

[4] https://en.wikipedia.org/wiki/XY_sex-determination_system

What you shall witness in this small book is the pattern of a bio-physical record how God began his self-creation in the womb of space. But were it not for the female factors involved, belonging to His Being, there is no Him.

Question: What formation of science did God leave in the DNA code to allow us to study His own self-creation?

Answer: A bio-chemical language comprising of four letters: **A** (Adenine), **T** (Thymine), **C** (Cytosine), and **G** (Guanosine); three nucleotides i.e., Hydrogen, 5 Carbon sugars, Phosphate and one source of energy, light.

These forces are not formless, they represent a part of His Being. His thoughts willed His own self-creation. He is the Originator; there was no pattern for Him to follow in what many consider His beginning. Therefore, none is like unto Him. He neither begets nor was He begotten; He was self-created without mother nor father. He was the first atom of life with weight. He was the first and last Black Being in space building self out of darkness into an orb (cell) of light. He lives forever because His thought i.e., replication process of self has never died.

> *"...Rasul Muhammad, had learned from a conversation he... had with Sister Burnsteen Mohammed, who served as secretary to Master Fard Muhammad in the early 1930s in Detroit, Michigan, how the Black Man traveled in the dark, vast regions of space...my son, he explained that this question was raised during a question and answer period [in 1930]...The questioner wanted to know how the Black Man traveled in the darkness of space before there was Sun, Moon or Stars and planets... Master Fard Muhammad, explained that the Black Man traveled in and out of "orbs."* [5]

---

[5] http://www.finalcall.com/artman/publish/Columns_4/article_101359.shtml

God's Genetic Pattern In Man and Woman

In a few words, all creation began with God's thought process or genetic code. Operating within human nature (restorative power) is a pattern of life replicating self-creation and self-organization. Our life is governed by His genetic code comprising chemical formula; fueled by energy, created seemingly out of nothing. Did nothing produce an atom, DNA, proteins, cells, tissues and organs such as heart, lung, liver etc.? Can nothing leave a pattern? What is a pattern? Patterns are applied mathematical calculations. A pattern is by definition a "Template" or *something that serves as a model for others to copy.* As such, the elements of a pattern repeat in a predictable manner. A geometric pattern is a kind of pattern formed of geometric shapes and typically repeating itself.[6] In our case, both male and female are patterned after, the first and last, Originator – God, who created all things, seemingly out of nothing.

The Honorable Elijah Muhammad taught that God is self-created out of the womb of blackness of space.

"I'm quoting just a few words from the Honorable Elijah Muhammad of which this involves the Honorable Minister Louis Farrakhan. This is taken from one of his books we can read this: "How came the Black God, Mr. Muhammad? He is Self-created.

"How could Self create Self? Take your magnifying glass and start looking at these little atoms out here in front of you. You see that they are egg-shaped and they are oblong. You crack them open and

---

[6] A **geometric pattern** is a kind of pattern formed of geometric shapes and typically repeated

you will find everything in them that you find out here. Then were there some of them (atoms) out here?

"Well who created them [atoms]? I want you to accept the Black God..."[7]

Question: If everything in an (Atom) we find today is out here, what do we see all around us, out here, today?

Answer: We see the creation of what is governed by applied mathematical calculations i.e., the sun, moon, stars, universe and earth. We see all manner of men and women, buildings, towns and cities, expressways, factories, transportation systems, electricity, computers etc., built by men and women. We see civilizations.

"...There was no light anywhere. Out of the total orbit of the Universe of darkness there sparkled an atom of life...

"How long was that? We can't tell; we weren't there. He was the only One in the whole entire dark Universe. He had to wait until the atom of life produced brains to think what He needed..."[8]

Question: What template of brains did the first atom of life produce as a pattern for us to study today?

Answer: We can study ourselves. We can study the brain and the pattern of its physical creation through today's science.

For instance, our physical brain, during human development, form in layers of three. (See Appendix 1) Today's scientist refer to brain cells, as neurons—an electrically excitable cell that receives, processes, and transmits information through electrical (light) and chemical

---

[7] http://www.finalcall.com/artman/publish/Columns_4/article_103254.shtml

[8] http://www.finalcall.com/artman/publish/Columns_4/article_103254.shtml

signals.[9] In principle, our brain corresponds with our body like to sun is to our solar system. They function without you giving it a thought. For instance,

a) Brain functions to maintain command over our bodies 9 major involuntary systems
b) The sun governs the major 9 planets of our solar system

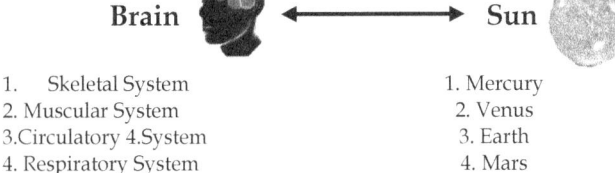

Brain	Sun
1. Skeletal System	1. Mercury
2. Muscular System	2. Venus
3. Circulatory 4.System	3. Earth
4. Respiratory System	4. Mars
5. Nervous System	5. Jupiter
6. Endocrine System	6. Saturn
7. Urinary System	7. Uranus
8. Digestive System	8. Neptune
9. Reproductive System	9. Pluto

All of the above systems are energized by light. What is light? Light is produced by the acceleration of charged particles and from the law of electromagnetism that states: An accelerated charge produce electromagnetic waves. Simply put, light is the transfer of energy from one part of electromagnetic field to another.[10] Therefore, as sun light strikes all 9 planets to keep each one alive, rotating and energized, so do brain cell neurons light up when sending information to our 9 involuatry systems keeping them energized and alive. Diagram 2 depicts brain cell neurons

---

[9] https://en.wikipedia.org/wiki/Neuron

[10] https://physics.stackexchange.com/questions/41680/why-is-light-called-an-electromagnetic-wave-if-its-neither-electric-nor-magne

lighting up as information is sent from one end of the brain to the other. Cell neurons are orbs of light, centered with a nucleus, communicating God's instructions.

Diagram 2

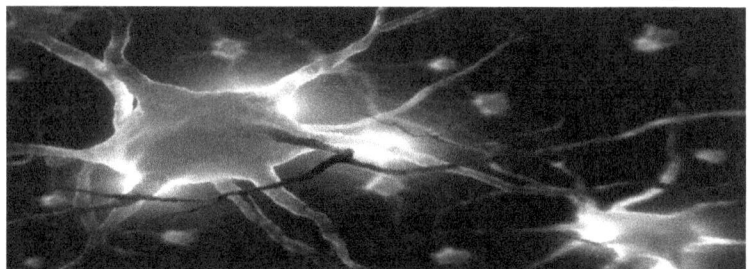

Bio-chemical interactions on [the skin of the brain cell] is the reason it generates a charge of light or electricity (fire). Just to name a few, chemicals involved, potassium (K^+), Sodium (Na^+) and chloride (Cl^-) lend to cellular voltage (light) provided there is atomic motion. In other words, life and light can only exist as long as creation is thinking. No thinking (or movement), no light. To human cells, light is its energy source; to celestial planets, light is its energy source. To the Originator of Self, light is a power He created out of nothing from His own thought and no one else's.

So, the Arabic word *"beda'a"* means creation of something out of nothing. This word also connotes the fact that something is created not on a pattern previously designed of something but as a completely new entity having no precedence. GOD IS A SCIENTIST, the First and the Last! He simply says, be and it is, over time. You ask, so how did God think of what it took to create the chemicals (atoms and ions) and elements to produce self out of nothing? That a good question.

Nature of Human Being Is Righteousness

The template and pattern of lights nature is yet operational within the nature of man and woman. Fact is, light accompanies the nature of every atom cell within the human being. Light energizes all created things at right angles. (See Diagram 3)

Diagram 3

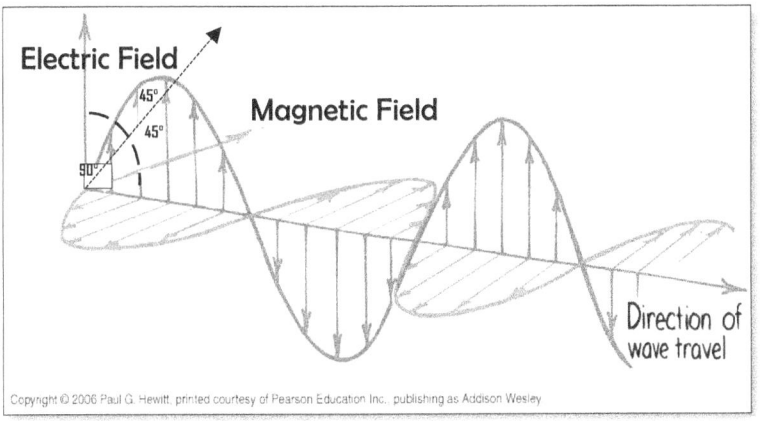

Diagram 3 demonstrates the properties of light in motion as a wave. Its electric *power* moves vertically. Its magnetic *force* moves horizontally which means both properties intersect at 45 degree right angles, an angle formed on a perpendicular intersection of **90°**. Light is created to propagate (spread messages of information to every cell), energy at 45 degree right angles. A moving power and force generates electricity and magnetism.

Since human beings consist of billions of cells, we are self-possessed with light too, at right angles to think right, move right and build right. So why do we live wrong, act wrong and think wrong? Answers to these questions shall be presented later. For now, let's merely say spiritual leaders of a community or nation represent the nucleus of the mind of the people's cultural psychology[11] as does the nucleus of an atom represent every aspect of its command center to express God's genetic code to build self. Essentially, men and women are physical beings governed by Gods' light under the process of His genetic code. Therefore, we are born, by nature, to be right! Doing right is a cultural issue.

Cultural Psychology Of God

The Christian Bible scarcely describes civilization on earth before the fall of civilization under the Caucasians authority. All the book sparsely describes is that an area of the east (East Africa, Egypt, Arabian Peninsula, and Mideast) was once a Garden of Eden full of trees, good food, gold, minerals, great rivers flowing and good people.

> *"A river watering the garden flowed from Eden; from there it was separated into four headwaters. [11] The name of the first is the Pishon; it winds through the entire land of Havilah, where there is gold. [12] (The gold of that land is good; aromatic resin and onyx are also there.)[13] The name of the second river is the Gihon; it winds through the entire land of Cush. [14] The name of the third river is the Tigris; it runs along the east side of Ashur. And the fourth river is the Euphrates."* (Genesis 2:10-14)

Furthermore, the Christian Bible nor the Holy Quran say how many years the righteous people of the east lived under a righteous cultural psychology before being deceived out of their own land and way of life. A way of life in accord with the nature in which light propagates. I mean, God

---

[11] Cultural psychology is the study of how **psychological** and behavioral tendencies are rooted in and embodied in culture. The main tenet of cultural psychology is that mind and culture are inseparable and mutually constitutive, meaning that people are shaped by their culture and their culture is also shaped by them. (source:wikipedia.org/wiki/Cultural_psychology)

consciousness was the cultural psychology and not the Caucasians professed God consciousness or lack thereof.

What is Cultural psychology? It is the study of how psychological and behavioral tendencies are rooted in and embodied in culture. Culture represents the customs, arts, social institutions, and achievements of a particular nation, people, or other social groups. The main tenet of cultural psychology is that mind and culture are inseparable and mutually constitutive. Meaning, people are shaped by their culture and their culture is also shaped by them through their collective power to establish or give organized existence to their own way of life.

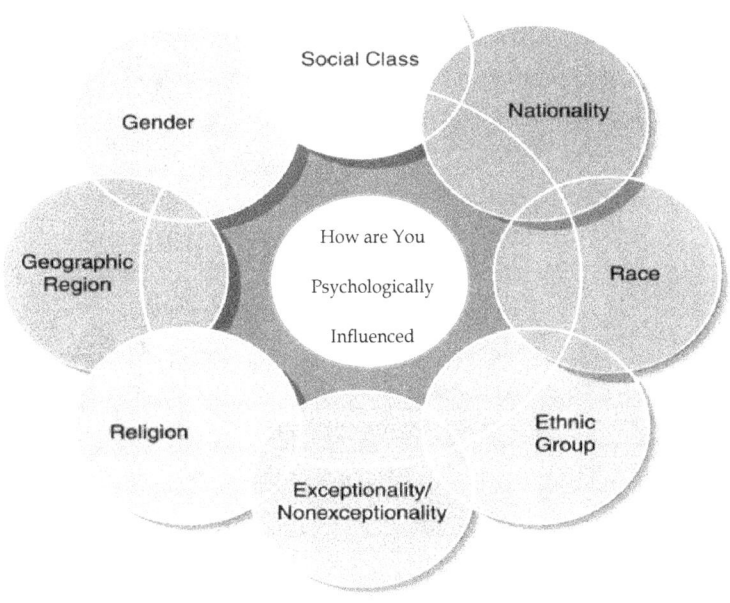

14

The question becomes: What is God's cultural psychology? Does He genetically extend His functionality into light to civilization through men and women?

Pre-Adamite civilizations built many magnificent cities and economies. Caucasians did not control world affairs back then. Right principles governed the cultural psychology of men and women. Right leadership aimed toward building civilizations using calculations, applications of geometrical principles and high science. It was a time when women were truly taught like mothers of civilization too. (See Ancient City Ruin Images beyond 4,000 years ago up to 12,000 years)

Ancient Cambodia City Ruins

Ancient Inca City Ruins

Ancient Kushite/Nubian City Ruins

Ancient South American City Ruins

Ancient Indus Valley City Ruins

Ancient Mexican City Ruins

Underwater Ancient City off Coast of Japan

Underwater City/Pyramids off coast of Coast Japan

Underwater Ancient Pyramids off Coast of Cuba

Of course there are many more such ancient ruins abandoned by members of the Original Black Nation all over the planet earth and buried by time. How they built it is one question. Why they built it is what todays black man and women of America must come know.

"According to the Honorable Elijah Muhammad,…we the Original people of the earth knew that an enemy was coming who would rule, we then set about to hide the wisdom of our ancient civilizations so that when the wicked one came, he could not borrow from our ancient wisdom to build his world, he would have to build it from himself," said Min. Farrakhan.

"These ancient civilizations of the Original, Black and Indigenous people were buried under jungles, sand and water and the only remnants that reminded the new man (Caucasian) that there was civilization before him were the pyramids and sphinx in Egypt…

"Scientists and archeologists have unearthed ancient cities and remnants of old civilizations in Asia, Africa, Central and South America and parts of the Middle East. Several years ago in the southeastern part of Turkey, remains of a 12,000 year old civilization was discovered under sand.

"Not only were we then to hide our wisdom and civilization, the woman who carries the secret of God would have to be called in to be protected from this new creature that was coming who would in his heart and mind, seek to destroy the moral character of women all over the earth. Now who are you that God would have to hide you and call you in to protect you because in you is the secret of God and civilization?"[12]

Question: What type of women walked the earth when our ancient civilizations were founded and flourishing compared to today's women?

Answer: Women who understood they were the mothers of civilization living among men who understood she is the first teacher of children.

Overall, our ancient communities were at liberty to act upon a cultural psychology to reflect their own nature and

---

[12] http://www.finalcall.com/artman/publish/National_News_2/article_102283.shtml

understanding of God's instinctive nature (restorative powers) to propagate right conduct, behavior and tendencies among society. Not a holy-perfect culture, but one wherein life was sacred. Decent social norms kept people leaning toward uprightness or suffer swift consequences.

To say the least, sacred world view communities were mutually constitutive[13] before the devil tribe ("deshru.t" – red ones (f) — white or light-skinned people)[14] entered the holy land. *(Read Holy Quran 7:27)* Since that time, a disruptive interference between good and evil has reached a feverish cultural battle. So now over the past 6,600 years or more, righteous cultural psychology is topsy-turvy. Murder, willful ignorance and perversion are social norms. As a result, society is wobbling in legal fiction, sport and play.

The Wobble In Nature

The Honorable Elijah Muhammad, according to his Minister, Louis Farrakhan, said the reason we commit acts of unrighteous and rebellion is due to a wobble in nature.

"The Honorable Elijah Muhammad taught us that nothing in the universe is "perfectly round." All of the planets are oblong; all of the stars are oblong. There is no "perfectly round" fruit; nor are there any heads of a human being that are "round" – our heads are shaped like the Earth, egg-shaped, oblong. And because The First Law of The Universe is Motion: Then anything that is not

---

[13] Both parties (male and female) will have an equal share and responsibility in creating or organizing something.

[14] https://www.africaresource.com/rasta/sesostris-the-great-the-egyptian-hercules/khememu-the-black-people-of-ancient-egypt/

perfectly round, and is set into motion, will have a "wobble" to its motion...

"In human nature, there is also a "wobble" and an imbalance, and an imperfection; this is why the Bible and Holy Qur'an agree that Allah (God) created "the heavens and the earth" in "six days" or "six periods of time"[15]

To understand the wobble in human nature, we must study the pattern that The Originator left in the science of His creation. For example, the motion of the earth has a wobble. (See Appendix 2)

Likewise our genetic codon processes wobble on RNA's 3rd base pair in the course of translating DNA. But before, RNA translates DNA, transcription must take place. You ask, what is transcription?[16] Transcription begins within chromosomal DNA's backbone. (See Appendix 3)

RNA arises from within the backbone of DNA before leaving to begin translating the genetic information. To properly understand this process in our nature is to perceive masculine and feminine forces at work. See Diagram 4, steps 1 and 2 for more details.

RNA strand inside DNA

---

[15] Farrakhan Speaks: Why God Made the Devil

[16] Transcription is the first step of gene expression, in which a particular segment of DNA is copied into RNA

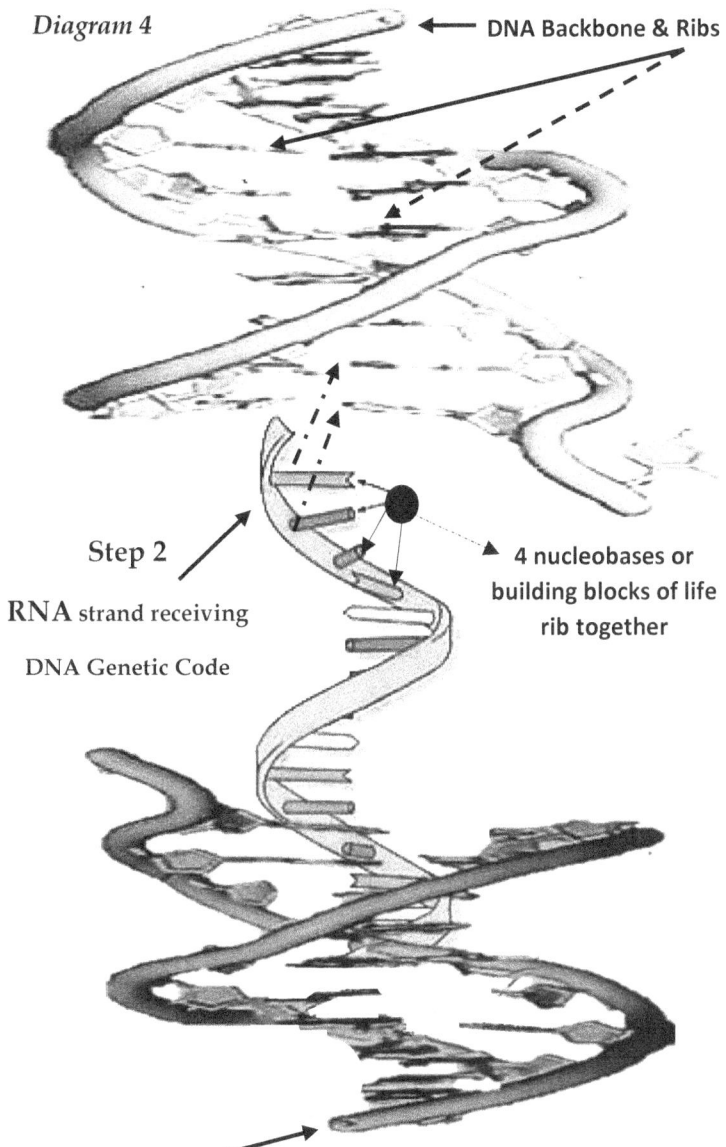

In between the backbone of the DNA are four (4)[17] foundational *(basic building blocks)* or nucleobases ribbed together in units of 2 base pairs.[18] The four DNA nucleobase bio-chemicals function as (Adenine-*A*) pairs with (Thymine-*T*) and (Guanine- *G*) pairs with (Cytosine- *C*). These 4 *basic building blocks* always match or remain faithful during DNA replication to make its second self or other half.

> *"The regular structure and data redundancy provided by the DNA double helix make DNA well suited to the storage of genetic information, while base-pairing between DNA and incoming nucleotides provide the mechanism through which DNA… replicates [a twin half], and RNA transcribes DNA into RNA.[19]*

This creative process begins from a single cell, thus the Holy Quran reads, *"It is He Who created you from a single being"…* a single cell of chromatin that is named for the material from which our DNA chromosomes make you into self. (See Diagram 5) What you may notice in diagram 5 is how a single human chromatid is patterned after the movement of the Analemma sun across are skies. How is that? (See Appendix 4)

---

[17] "In **FOUR Days** He placed the mountains on it blessed it, and equally measured out sustenance for those who seek sustenance." (Holy Quran 41:10)

[18] "So He completed them as seven firmaments in **TWO Days**, and He assigned to each heaven its duty and command. And We adorned the lower heaven with lights, and (provided it) with guard. Such is the Decree of (Him) the Exalted in Might, Full of Knowledge." (Holy Quran 41:9-12)

[19] https://en.wikipedia.org/wiki/Base_pair

Diagram 5

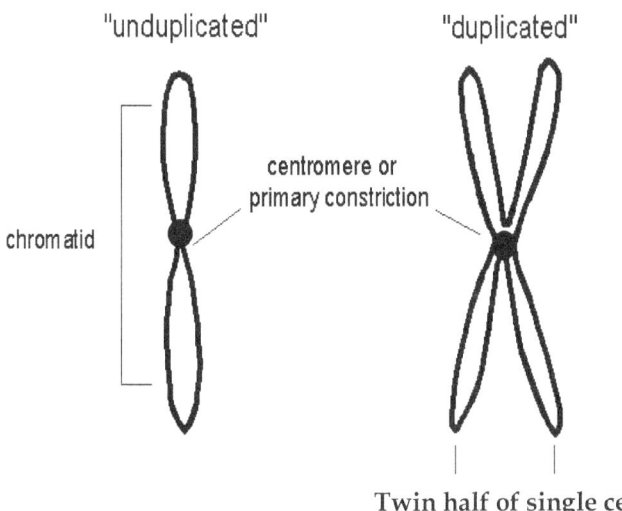

Twin half of single cell

*Say: Travel **in the earth then see how** He makes **the first creation, then Allah creates the latter creation. Surely Allah is Possessor of power over** all things. **Surely Allah is Possessor of power over** all things. **(Holy Quran 29:20)***

To recap, DNA genetic information is the original language of God to produce both male and female. God's code of DNA is written with four letters *(biological compounds or nucleotides)* represented as **A - T, G - C, T - A** and **C – G.** These specific mixtures all contain Hydrogen, Nitrogen, and 5 Carbon Sugars. (See Diagram 6)

Diagram 6

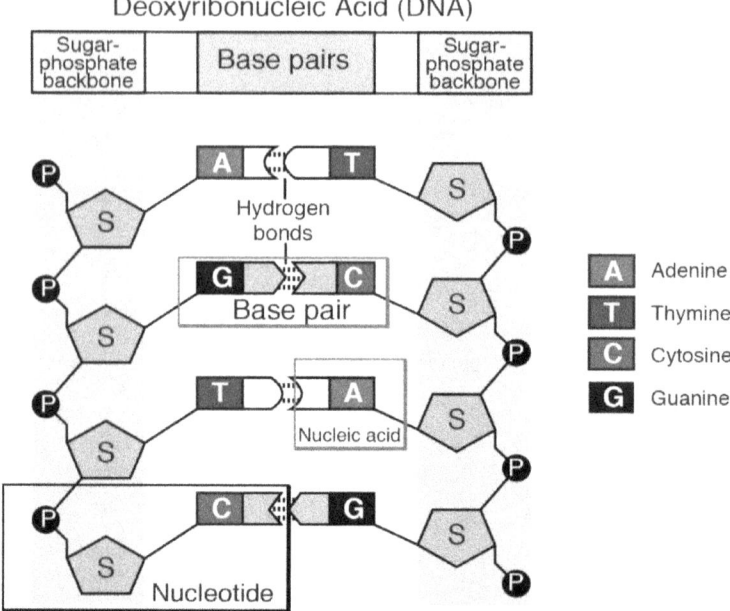

While pairing, these four subterranean building blocks of life are guarded during DNA replication because they do not mismatch. But if they do, an editing process make corrections to assure the twin half of the single cell is mated correctly like a carbon copy. (See Diagram 7)

> *"Transcription has some proofreading mechanisms, but they are fewer and less effective than the controls for copying DNA. As a result, transcription has a lower copying fidelity than DNA replication."*[20]

---

[20] https://en.wikipedia.org/wiki/Transcription_(biology)

Diagram 7

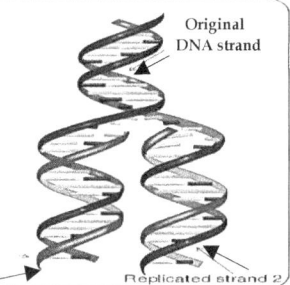

DNA Replication
- Prior to cell division DNA has to make a copy of itself= **DNA Replication**
- Follows a template mechanism for this process much like a photograph is made from a photo negative.
- More than a dozen enzymes are involved in the process.

Original DNA strand
Replicated DNA Strand 1
Replicated strand 2

As mentioned earlier, after RNA leaves the DNA nucleus with its genetic code, the wobble happens on its *secondary* strands 3rd base pair position, but why? At this juncture, RNA code switches one biological compound called uracil. Uracil (I-U) is unfaithful and therefore bonds with multiple nucleobase pairs i.e.:

1. guanine-uracil (G-U)
2. hypoxanthine-uracil (I-U)
3. hypoxanthine adenine (I-A)
4. hypoxanthine-cytosine (I-C)

"Pairing of the secondary structure of RNA during translation anticodon with the tRNA codon proceeds from the 5' end of the codon. Once the first two positions are paired, exact base pairing of the third position is less critical.

*"The third (5') base of the anticodon can typically pair with either member of [DNA's letters] A, T, C, and G in the codon as appropriate: it "wobbles"... This allows mRNA to be translated with fewer than the 64 codon's for **20 amino acids** that would be required without wobble. Some wobble positions can pair with any of the four bases."*[21]

---

[21] https://www.mun.ca/biology/scarr/MGA2_03-22.html

Continue to view Diagram 8 below and follow steps 1 – 5 for basic understanding RNA wobble pairing.

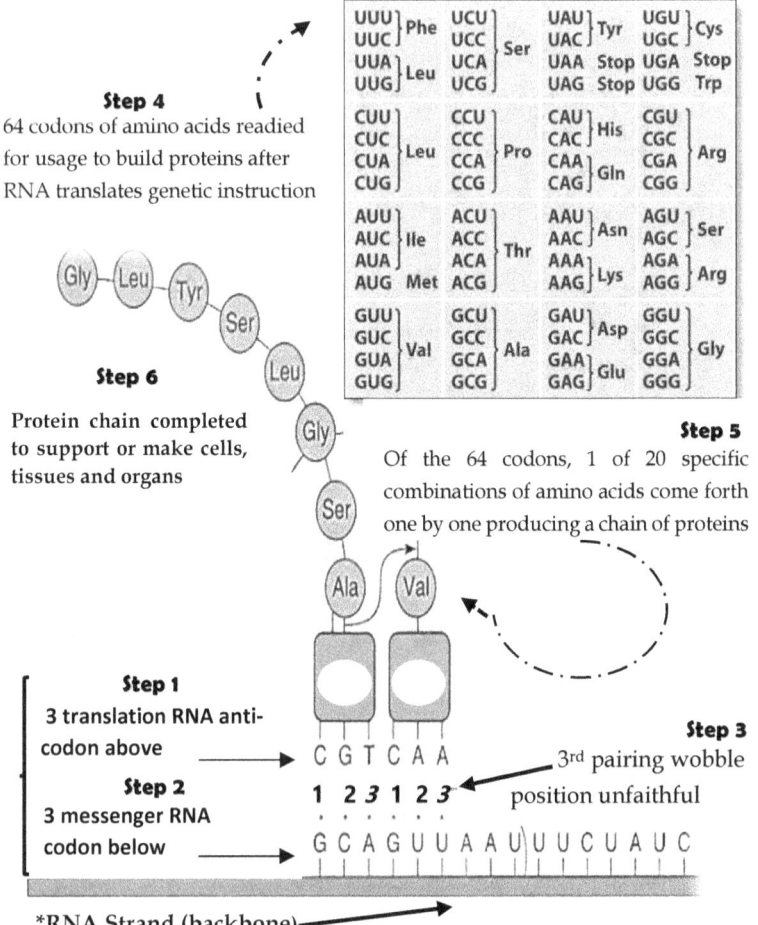

Step 4
64 codons of amino acids readied for usage to build proteins after RNA translates genetic instruction

Step 6
Protein chain completed to support or make cells, tissues and organs

Step 5
Of the 64 codons, 1 of 20 specific combinations of amino acids come forth one by one producing a chain of proteins

Step 1
3 translation RNA anti-codon above

Step 2
3 messenger RNA codon below

Step 3
3rd pairing wobble position unfaithful

*RNA Strand (backbone)

> "[RNA's] wobble ***effect***. If the point mutation occurs in the third nitrogen base in a codon, then it causes no effect on the amino acid or subsequent protein and the mutation does not change the organism. At most, a point mutation will cause a single amino acid in a protein to change.
>
> "While this usually is not a deadly mutation, it can cause issues with that protein's folding pattern and the tertiary and quaternary structures of the protein."[22]

RNA's capability to make proteins for every kind of cell, tissue or organ throughout our body is a masculine and feminine quality. The end product of proteins are like offspring. Once proteins are formed, they only exist for a certain period and are then degraded and recycled by the cell's machinery through the process of protein turnover.[23]

Again, RNA translates genetic codon once it is outside DNA's nucleus cell. Then the wobble effect supervenes opening a doorway for mutations. This wobble in the subterranean genetic codon of creation also extends into our personal issues and the civilizations we have built, hence they are collapsible as this present western world order is also folding up due to the wobble.[24] None can escape its effect until a new and perfect creation is brought about. Read Holy Quran 40:57 and Revelations 5:21 to grasp why the Originators first creation was not perfect, yet it was created to move toward perfection, stage after stage.

---

[22] https://www.thoughtco.com/dna-mutations-1224595

[23] https://en.wikipedia.org/wiki/Protein

[24] https://www.nobelprize.org/educational/medicine/dna/a/translation/trna_wobble.html

Number 8

So far, what the reader should be are witnessing in this small book is the aftermath of a productive process that happened trillions of years ago. Amazingly, the pattern of God's own self-creation has been recorded, translated and replicated throughout the production of time. A single "Being" thought and then protected Himself with a black endoderm of melanin, trillions of years ago. He is the first single cell or first atom of life. This Black God of space, time and energy lives in both male and female today. His biochemical compounds, the basic building blocks of that is! His nature is our nature (restorative powers). Both male and female are patterned after His Original subterranean biochemical genetic code. The Originator's (*beda'a* of self-creation) "proper" name is Allah! *"It is He Who created you from a single being, and then from the same being created its spouse, and sent down for you eight pairs of animals; He creates you in your mothers wombs, from one sort to another, in a triple darkness; such is Allah, your Lord for Him only is the kingship; there is no God except Him; so where are you being turned away?" (Holy Quran 39:6)*

To end this point on another scientific observation concerning an Atom, the Quran also says, *"And the angels will be on its sides, and (8) eight angels will, that Day, bear the Throne of your Lord above them. That Day shall you be brought to Judgement, not a secret of you will be hidden. Then as for him who will be given his Record in his right hand will say: "Take, read my Record!" (Holy Quran 17:19)* Why did Allah make mention of (8) pairs of animals? Here it represents electrons that provide stability to strengthen His own self-creation. Accordingly, all atoms follow that pattern seeking 8 electrons around its outer shell for stability and bonding strength.

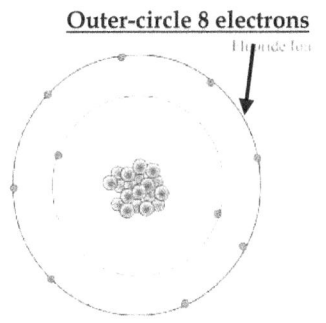

- Atoms are stable when they have 8 valence electrons.
- When the atoms have 8 electrons, it is called an octet.
- Atoms must lose, gain or share electrons to attain the octet.
- Atoms that form bonds with other atoms by sharing them or transferring them is known as the Octet Rule.

"Atoms like to be stable. They go to great lengths to be stable and happy, just like humans. Stability for an atom means having a complete outer energy level full of valence [the combining power of an element] electrons. The number of [valence electrons is what determines the properties of an element, especially whether or not it is stable. The Noble Gases have 8 electrons in the valence shell, which is why they don't react with any other element... ever."[25]

God Speaks With Numbers

Every atom created was created based upon God's original mathematical calculations. Atoms carry weight,

---

[25] www.quora.com/What-do-atoms-have-to-do-to-get-8-electrons-in-their-valence-shells

mass and emit light. For example, the chart below provides samples of atomic numbers, symbols, name of atom and their weight.

Atomic #	Symbol	Name	Atomic Weight
1	H	Hydrogen	1.008
2	He	Helium	4.002 602(2)
3	Li	Lithium	6.94
4	Be	Beryllium	9.012 1831(5)

The numerical value of each atom distinguishes its uniqueness and function. These unseen atoms all around us are not formless and neither is God. He became anthropomorphic. To build self-up physically, He created 92 natural elements containing specific numbers of protons and neutrons within the nucleus of each atomic element.[26] Do these elements contain living DNA? No. Yet they are atoms used in various combinations to physically transform into **anthropomorphic** body, from thought (gene expression) to light, to nucleobase, to single cell, the tissue to organs. Ninety-two elements employed to make physical plants, stars plant life, sea life, air, fire and water. (See Diagram 9)

---

[26] www.bing.com/search?q=define+atomic+mass

Diagram 9

Natural Elements and Symbols

Actinium	Ac	Europlum	Eu	Molybdenum	Mo	Scandium	Sc
Aluminum	Al	Fluorine	F	Neodymium	Nd	Selenium	Se
Antimony	Sb	Francium	Fr	Neon	Ne	Silicon	Si
Argon	Ar	Gadolinium	Gd	Nickel	Ni	Silver	Ag
Arsenic	As	Gallium	Ga	Niobium	Nb	Sodium	Na
Astatine	At	Germanium	Ge	Nitrogen	N	Strontium	Sr
Barium	Ba	Gold	Au	Osmium	Os	Sulfur	S
Beryllium	Be	Hafnium	Hf	Oxygen	O	Tantalum	Ta
Bismuth	Bi	Helium	He	Palladium	Pd	Tellurium	Te
Boron	B	Hydrogen	H	Phosphorus	P	Terbium	Tb
Bromine	Br	Indium	In	Platinum	Pt	Thorium	Th
Cadmium	Cd	Iodine	I	Polonium	Po	Thallium	Tl
Calcium	Ca	Iridium	Ir	Potassium	K	Tin	Sn
Carbon	C	Iron	Fe	Promethium	Pm	Titanium	Ti
Cerium	Ce	Krypton	Kr	Protactinium	Pa	Tungsten	W
Cesium	Cs	Lanthanum	La	Radium	Ra	Uranium	U
Chlorine	Cl	Lead	Pb	Radon	Rn	Vanadium	V
Chromium	Cr	Lithium	Li	Rhenium	Re	Xenon	Xe
Cobalt	Co	Lutetium	Lu	Rhodium	Rh	Ytterbium	Yb
Copper	Cu	Magnesium	Mg	Rubidium	Rb	Yttrium	Y
Dysprosium	Dy	Manganese	Mn	Ruthenium	Ru	Zinc	Zn
Erbium	Er	Mercury	Hg	Samarium	Sm	Zirconium	Zr

Simply put, Allah comprises in some shape, fashion and numerical formula: (a) thought, (b) light, (c) 4 nucleobase pairs, including a wobble effect (d) combination of 20 amino acids out of 64, and (e) 92 elements in various formations. All of these sources represent extensions of Allah's attribute *Al-Muhsee:* The Counter, The Reckoner, The One who the count of things are known to him. Both male and female are merely a part of Him by nature!

Let There Be Light

Man and women are biological compounds energized by light. Today's modern scientist refer to this type of light as bio-photons.

> *"In the 1970s Fritz Popp and a team of researchers at the University of Marburg started doing work with biophotons. Biophotons are considered ultra-weak photo emissions (UPEs). Popp's work has transformed our understanding of biophotons and the role they play. At one point biophotons were considered byproducts of chemical reactions within our DNA. We now know that the biphotons emitted from our cells are highly coherent [organized] energy that may be responsible for the operation of our biological systems.*
>
> *"What does this mean? We know that our cells emit light and that this light is constantly sending and receiving information."* [27]

Why did the angel Jabril (Gabriel) reveal to Prophet Mohammed of Arabia, Ibn Abdullah of 1,400 years ago that Allah is light? *"Allah is the Light of the heavens and the earth. The similitude of His light is as a niche wherein is a lamp. The lamp is in a glass. The glass is as it were a glittering star, lit from a blessed olive-tree, neither of the East nor of the West, whose oil would almost glow forth (of itself) though no fire touched it. Light upon light, Allah guides unto His Light whom He pleases, and Allah sets forth similitudes for mankind, and Allah is All-Aware of all things."* (Holy Quran 24:35) This God is not a deity. His name is not Allah, but Allah is His Title! He is depicted as a physical living being with internal organs and infinite wisdom able to generate light. The bible says, 1 John 1:5, *"This*

---

[27] www.drjoedispenza.com/blog/health/biophotons-the-light-in-our-cells/

is the message we have heard from him and declare to you: God is light; in him there is no darkness at all."

There are seven (7) forms of light (electromagnetic radiation) being emitted from the sun. We can say, all seven operate deep within the first atom cell of life as a source of possible energy and they are physical particles; not formless particles. Our eye cell rods and cones[28] can see only one form of the seven particles of light. The other six (6) forms, our eye cells do not detect. The range of colors of all seven forms of light is red, orange, yellow, green, blue, indigo, and violet. These represent various frequencies of light. When all seven forms of light merge together, that's the visible light our brain interprets from the eye cells as red, green, yellow, blue, etc., etc. (See Diagram 10)

Diagram 10

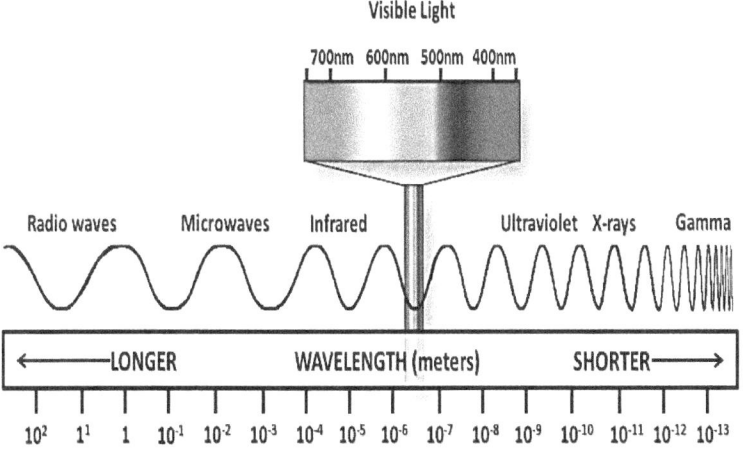

---

[28] www.physicsclassroom.com/class/light/Lesson-2/Visible-Light-and-the-Eye-s-Response

So rather we look into an atom, a human cell or the four basic chemical compounds of life, they are all composed of some form of light. I repeat, the atomic energy applied to move our subterranean self-creative processes from one stage to the next was/is LIGHT — electric magnetic radiation propagating information from Allah's thought at 45°right angles.

> "Indeed, the human body emits **biophotons**, also known as ultraweak photon emissions (UPE), with a visibility 1,000 times lower than the sensitivity of our naked eye. While not visible to us, these particles of light (or waves, depending on how you are measuring them) are part of the visible electromagnetic spectrum (380-780 nm) and are detectable via sophisticated modern instrumentation."[29]

Western civilization is designed to destroy human light. By turning civilizations cultural psychology into a satanic world of sport and play, drunkenness and sexual perversion, both male and female wobble through this life. Can we choose to be vehicles and conduits of light again?

> "AND SO, set thy whole being right towards the [one ever-true] faith, turning away from all that is false, in accordance with the natural disposition which God has instilled into man: [for,] not to allow any change to corrupt what God has thus created this is the [purpose of the one] ever-true faith; but most people know it not. (Asad Quran Translation Quran 30:30)

---

[29] Herbert Schwabl, Herbert Klima. **Spontaneous ultraweak photon emission from biological systems and the endogenous light field.** Forsch Komplementarmed Klass Naturheilkd. 2005 Apr; 12(2):84-9. PMID: 15947466 & Hugo J Niggli, Salvatore Tudisco, Giuseppe Privitera, Lee Ann Applegate, Agata Scordino, Franco Musumeci. **Laser-ultraviolet-A-induced ultraweak photon emission in mammalian cells.** J Biomed Opt. 2005 Mar-Apr; 10(2):024006. PMID: 15910080

Appendix 1
Brain Creation

On day one when sperm combines with the egg, it carries 23 chromosomes to mix with the 23 chromosomes in the egg. Sperm cells produce both x and y chromosomes and egg cells only produce x chromosomes. Therefore, the biological sex of the human being is determined by his y chromosome. In other words, if the sperm carries an x chromosome to mix with the egg, it's a female under normal circumstances. If the sperm carries a y chromosome, it's a male under normal circumstances. This self-creative process is under the guidance of the Originator whose name is spelled in Arabic letters, A L L A H. Coincidence or not, as the embryo moves in the placenta, its body spells God in arabic letters i.e., A aligns with the head/brain, L with the umiblical cord; the other L with the arms/hands and the H with the legs/feet.

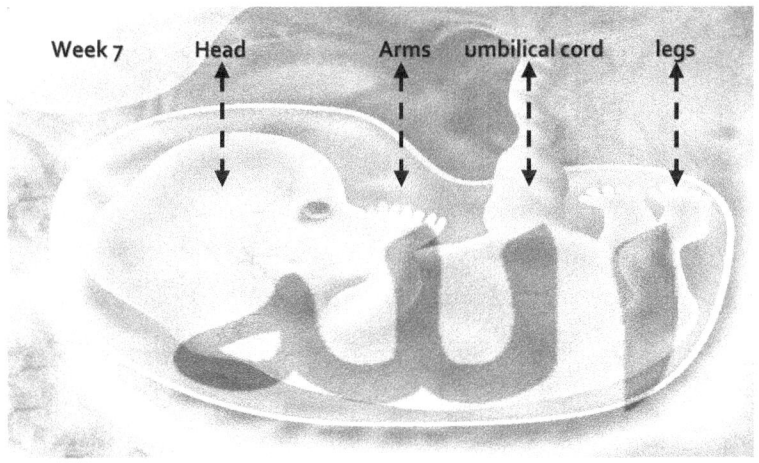

Lastly, as the brain begins to manifest in 3 layers of germ cells by about day 15, after fertilization, that is when the entire human body begins to take formation.

> "Stem cells have divided and differentiated into three different *germ layers* called 1) *ectoderm*, 2) *endoderm*, and 3) *mesoderm*. Each gives rise to major components of specific body structures and organs. Ectoderm derivatives include the skin, nails, hair follicles, sweat glands, and nerves within the lungs. **Another specialized layer of cells appearing at this time is the *neuroectoderm*, which gives rise to the brain, spinal cord, and peripheral nerves, as well as many of the muscles and bones in the face.**
>
> "Endoderm forms the lining of the respiratory and gastrointestinal tracts and gives rise to major portions of internal organs including the lungs, liver, pancreas, and intestines.
>
> "Mesoderm derivatives include the heart, kidneys, bones, muscles, and blood vessels as well as portions of the reproductive and urinary systems.
>
> "Mesoderm also gives rise to specialized cells called *somites* (so'mits). These cells form most of the skull and ribs as well as the vertebral column or backbone.
>
> "All of these cell layers and cell types work in concert forming the increasingly complex embryo."[30]

Be mindful here, the ectoderm consisted of the black skin (dominant melanin gene) because in the beginning, there were no recessive genes. Only dominant and incomplete dominant that generate black or reddish-brown pigmentation. **Hence, the original body formed by the brain is called "primitive streak"** to signify the creation of a unique, human being.[31] The primitive streak contains pure melanin (blackness). Essentially, all people originate from The Original Black God! His left his mark even on the so-called white race. He wears the title ALLAH!

---

[30] http://www.ehd.org/dev_article_unit3.php
[31] https://en.wikipedia.org/wiki/Primitive_streak#Ethical_implications

Appendix 2

NASA Solves Mysteries About Wobbling Earth

In the process of solving this recent mystery, the researchers unexpectedly came up with a promising new solution to a very old problem, as well. One particular wobble in Earth's rotation has perplexed scientists since observations began in 1899. Every six to 14 years, the spin axis wobbles about 20 to 60 inches (0.5 to 1.5 meters) either east or west of its general direction of drift."[32]

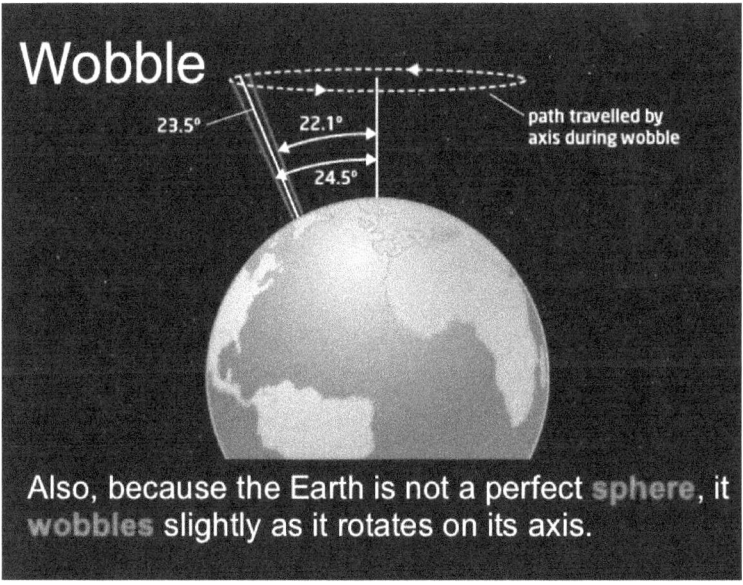

Why don't we feel the wobble of the earth? As the Honorable Minister Farrakhan has taught, the earth is under a law that it obeys according to its creator.

---

[32] https://www.jpl.nasa.gov/news/news.php?feature=6332

Appendix 3

DNA and RNA Masculine and Feminine Principles

And Jehovah Elohim built the rib that he had taken from Man into a woman; and brought her to Man. (Genesis 2:22)

If you are keen, notice the word rib in the scientific name DNA (***deoxyribose***), and in **RNA** (***ribose***). The difference here is that in deoxyribose (DNA) there is no -OH (Oxygen/Hydrogen) molecule present on the second carbon of the sugar molecule, but it is found in ribose (RNA).

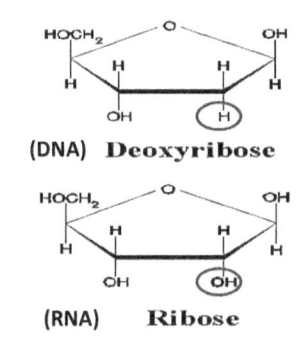

3 Differences Between DNA and RNA nucleotides:

1) sugar in RNA is ribose

2) 2' Carbon on Ribose has (OH group) and on Deoxyribose there is just a Hydrogen

3) DNA has the base thymine (T), RNA has the base **uracil (U)**

-OH groups are quite reactive and this makes RNA less stable than DNA, and more susceptible to degradation.[33] Due to the different features of DNA and RNA's nature, the wobble occurs on the 3rd base pair of RNA while translating DNA's genetic code into cells, tissues and human organs.

He was created from a fluid, ejected. Emerging from between the backbone and the ribs. (Quran 86:6-7)

---

[33] www.thenakedscientists.com/forum/index.php?topic=45740.0

Appendix 4
Analemma Sun Pattern

Image A (Suns pattern of movement from Earth single cell chromatid)

…from Earth's perspective the sun is constantly moving in a figure-of (8) eight pattern? It's called an analemma –that image (A) depicts the sun's movement in the sky over a period of one year, when observed from a fixed position on the Earth. This happens because of the Earth's axial tilt making it look like our sun is changing its position in the sky every day. You can easily do your own analemma. What you'll need is a tripod, a camera and quite a bit of patience. Just keep the camera at a fixed location and orientation and take multiple shots throughout the year, always at the same time of the day.

Oh, and make sure to scroll down to the bottom of the gallery where you'll find Lunar analemma and even one from Mars, half our chromosome shape!

Image B (Suns pattern of movement from Mars half cell chromatid)

Glossary

Gene: A gene is a locus (or region) of DNA that encodes a functional RNA or protein product, and is the molecular unit of heredity. The transmission of genes to an organism's offspring is the basis of the inheritance of phenotypic traits.

Germ Cell: Any biological cell that gives rise to the gametes [marriage] of an organism that reproduces sexually. In many animals, the germ cells originate in the primitive streak and migrate via the gut of an embryo to the developing gonads. There, they undergo meiosis, followed by cellular differentiation into mature gametes, either eggs or sperm.

Pattern: A pattern, apart from the term's use to mean "Template", is a discernible regularity in the world or in a manmade design. As such, the elements of a pattern repeat in a predictable manner. A geometric pattern is a kind of pattern formed of geometric shapes and typically repeating like a wallpaper.

Black God: Ancient Egyptian Hieroglyphics = *Kmt (Black) [34] wr (man, God, command)[35] *Kmt is a feminine noun as the "t" is silent in this ancient Egyptian expression.*

Neurons: Neurons are surrounded by a cell membrane. Neurons have a nucleus that contains genes. Neurons contain cytoplasm, mitochondria and other organelles. Neurons carry out basic cellular processes such as protein synthesis and energy production.

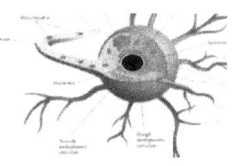

---

[34] https://www.africaresource.com/rasta/sesostris-the-great-the-egyptian-hercules/khememu-the-black-people-of-ancient-egypt/

[35] http://asarimhotep.com/documentdownloads/AfricanOriginsoftheWordGod.pdf

Caucasian: Originally referred in a narrow sense to the native inhabitants of the Caucasus region. The United States National Library of Medicine often used the term "Caucasian" as a race in the past. However, it later discontinued such usage in favor of the more narrow geographical term *European*, which traditionally only applied to a subset of Caucasoids.

Legal Fiction: An assumption made by a court and embodied in various legal doctrines that a fact or concept is true when in actuality it is not true, or when it is likely to be equally false and true. A legal fiction is created for the purpose of promoting the ends of justice.

Wobble: Move unsteadily from side to side.

Nitrogenous base: Organic molecule with a nitrogen atom that has the chemical properties of a base. The main biological function of a nitrogenous base is to bond nucleic acids together. A nitrogenous base owes its basic properties to the lone pair of electrons of a nitrogen atom.

Subterranean: Life existing, things occurring, or done under the earth's surface or atomic cellular life.

Geometry: Branch of mathematics concerned with questions of shape, size, relative position of figures, and the properties of space.

Chromosome: A packaged and organized structure containing most of the DNA of a living organism. It is not usually found on its own, but rather is structured by being wrapped around protein complexes called nucleosomes, which consist of proteins called histones.

RNA: Ribonucleic acid (RNA) is a polymeric molecule implicated in various biological roles in coding, decoding, regulation, and expression of genes. RNA and DNA are nucleic acids, and, along with proteins and carbohydrates, constitute the three major macromolecules essential for all known forms of life.

DNA: Deoxyribonucleic acid is a molecule composed of two chains which coil around each other to form a double helix carrying the genetic instructions used in the growth, development, functioning and reproduction of all known living organisms and many viruses. DNA and ribonucleic acid are nucleic acids; alongside proteins, lipids and complex carbohydrates, nucleic acids are one of the four major types of macromolecules that are essential for all known forms of life.

Chemical compound: Any substance composed of identical molecules consisting of atoms of two or more chemical elements. All the matter in the universe is composed of the atoms.

Valence: Electron that is associated with an atom, and that can participate in the formation of a chemical bond; in a single covalent bond, both atoms in the bond contribute one valence electron in order to form a shared pair to strengthen with energy.

Biophotons: Photons of light in the ultraviolet and low visible light range that are produced by a biological system. They are non-thermal in origin, and the emission of biophotons is technically a type of bioluminescence, though bioluminescence is generally reserved for higher luminance luciferase systems.

Anthropomorpic: The attribution of human traits, emotions, or intentions to non-human entities. It is considered to be an innate tendency of human psychology.

Adamic Race: You have the Adamic race and you have the pre-Adamites. That means there were people on the earth before Adam. The pre-Adamites is the *us* [Black people] out of which Adam came...There were pre-Adamites on the earth before Adam. If Adam was only one person, Eve was one person, Cain was one person, Abel was one person, and, Cain slew Abel, then, Cain goes into the land of Nod and he finds himself a woman in the land of Nod, then, where did she come from? What family was she a part of? –Minister Louis Farrakhan--[36]

Pre-Adamite: Original ancient Black and Brown civilizations dating back prior to 6,000 years ago.

Thought: Travels 24 billion miles per second.[37]

Homeostasis: The property of a system in which variables are regulated so that internal conditions remain stable and relatively constant. Examples of homeostasis include the regulation of temperature and the balance between acidity and alkalinity (pH) in water.[38]

---

[36] Adam and the pre-Adamites by The Honorable Minister Louis Farrakhan April 8, 2001 message delivered by Min. Louis Farrakhan at Mosque Maryam in Chicago

[37] Problem Book, Nation of Islam

[38] https://en.wikipedia.org/wiki/Homeostasis

DNA Attributes Are Attributes of Alllah

The word Allah is not the name of God. Allah is the Title of the Divine Supreme Being. His patent 99 attributes are characteristics or inherent traits extending into both male and female, including our DNA. For the purpose of this little 44 page book, I will only demonstrate a few examples shown below with respect how Allah is featured in our DNA:

DNA	ATTRIBUTE
Gene (genetic inheritance)	*al-Wārith* (inheritor)
TATA box (promote expression)	**al-Muqaddim** (the promoter)
Chromosome (instructions)	*al-Muṣawwir* (incline to dialog)
Bio Photon (light)	*an Nur* (to illuminate)
Atom proton/neutron (binding force)	*al ahad* (connect, join, and unite)
Nucleic Acid (life building blocks)	*al-Bāri'* (create, evolve)
Oocyte (ova genesis)	*al-Mu'akhir* (the delayer)
Gilal (brain homeostasis)	*al-Muqeet* (guardian, maintainer)

A final remark with respect to the attribute *al-Mu'akhir* (The Delayer) is that it extends into the oocyte in the creation of female eggs. Female eggs form at birth via embryotic stage. Then its development is DELAYED for years until the female begins her menstrual cycle.[39] What a wonderful and mighty God we have within. The question is: Can we align our free will (faculty of consciousness) with the faculty of the supers-consciousness of Allah who inhabits us while live? Since His DNA extends into our DNA, then who and where is God?

---

[39] https://embryology.med.unsw.edu.au/embryology/index.php/Oocyte_Development